Our Ever-Ch... Earth

Written by Rebecca Stark

ISBN 978-1-56644-052-3

Educational Books 'n' Bingo

Previously published by Educational Impressions, Inc.

Printed in the U.S.A.

Table of Contents

To the Teacher

Our Ever-Changing Earth is an informative, fun-filled unit about geology. The book explores the structure and composition of our planet Earth; rocks and minerals; mountain building; weathering, erosion, and sedimentation; plate tectonics; and the oceans. It also deals with earthquakes, volcanoes, and other geological hazards.

Students are presented with opportunities to practice crucial critical- and creative-thinking skills. A variety of types of activities are included: creative writing, research, analyzing, evaluating, and more. A fun What's the Question? game is provided to reinforce the concepts learned in the unit. In addition, a crossword puzzle is included. The puzzle may be used to evaluate knowledge or just for fun!

Geology—What Is It?

The word "geology" comes from two Greek words: *geo*, meaning "Earth," and *logos*, meaning "study." And that's what geology is—it's the study of Earth's history, composition, and structure as well as the processes which affect it. Scientists who study these things are called **geologists**.

Geology is divided into two main fields: physical geology and historical geology. **Physical geology** focuses on the composition of Earth, the movements that take place within and upon Earth's crust, and the processes which affect Earth's materials and structure. There are many subfields of physical geology. The main ones include **mineralogy, petrology, structural geology, plate tectonics, geomorphology, economic geology, geophysics,** and **geochemistry.** There are other, more specialized, subfields as well.

Historical geology focuses on the development of Earth. Historical geologists study rocks to learn about the evolution of the planet and its life forms. They divide geological time into four main intervals, or eras—the Precambrian Era, the Paleozoic Era, the Mesozoic Era, and the Cenozoic Era. Main subfields include **paleontology, stratigraphy,** and **paleogeography**.

"–Ologies"

Match the subfield of geology on the left with the description on the right. If your answers are correct, they should spell a synonym for "marine geology." (Note: A_1 must come before A_2 and O_1 before O_2.)

SUBFIELD

_____ 1. Mineralogy

_____ 2. Petrology

_____ 3. Structural geology

_____ 4. Geomorphology

_____ 5. Economic geology

_____ 6. Seismology

_____ 7. Marine geology

_____ 8. Volcanology

_____ 9. Paleontology

_____ 10. Geochronology

_____ 11. Paleoclimatology

_____ 12. Glaciology

THE STUDY OF...

A_1. General configuration of earth's surface and of landforms

N. Geologic processes and materials which can be utilized by humans

G. Ocean floor and ocean-continent margins

O_1. Classification of minerals

A_2. Fossils of prehistoric plants and animals

C. Origin, structure, occurrence and history of rocks

H. Climates of the geologic past

E. Deformation of rocks and their structural arrangements

P. Time in relation to the history of the earth

O_2. Earthquakes and the earth's interior

Y. Glaciers and glaciation

R. Volcanoes

Another word for "marine geology" is _____.

Earth's Structure and Composition

Like the other inner planets of our solar system, Earth is a rocky planet. Although we are able to drill only a few miles into the surface of the Earth, we still know a lot about its structure and composition. The Earth is layered into distinct concentric shells: the crust, the mantle, the outer core, and the inner core.

The **crust,** or outer shell, is relatively thin, ranging from about four miles to about twenty-five miles in thickness. There are two kinds of crust: the continental crust, which forms the land masses, and the thinner oceanic crust, which forms the seafloor.

Beneath the crust is the **mantle**. At about 1,800 miles thick, the mantle comprises most of Earth's mass and volume. The crust and the upper 60-mile layer of the mantle is called the lithosphere. The lithosphere is cool and strong. Beneath this cool, strong layer is the asthenosphere. Because the asthenosphere is warm, partial melting takes place in this zone, making it rather weak and prone to shearing by the rigid lithosphere. Shearing along this zone is what causes continental drift.

Beneath the mantle is a dense **core**. Geologists believe that the outer 1,380 miles of the core are molten and that the inner 780 miles are solid. It is the outer core that is responsible for Earth's magnetic field.

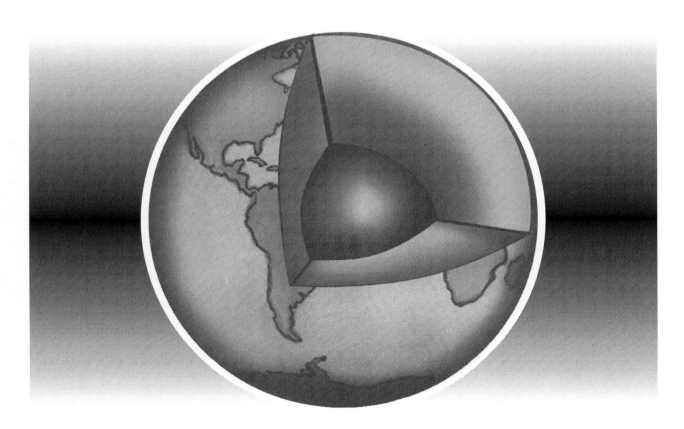

Earth Layers

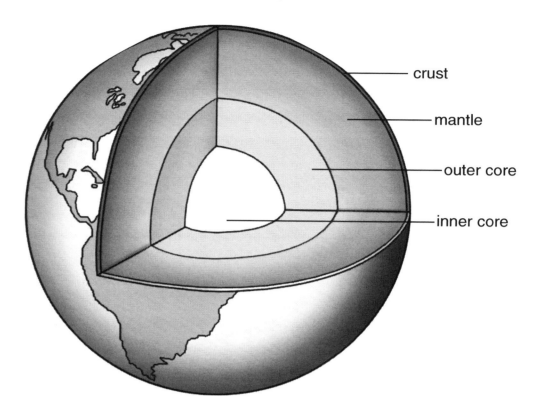

crust

mantle

outer core

inner core

Activities

A blanket of air surrounds our planet. Without this blanket of air, life as we know it would not be possible. Scientists divide the Earth's atmosphere into layers. Name those layers from lowest to highest.

Choose another planet in our solar system. Compare and contrast the structure of that planet with that of planet Earth.

Activities

Create a geology board game.

Research and find out what is meant by the Mohorovičić discontinuity, or Moho.

Make a flipbook of geological hazards.

Scientists believe Earth to be at least 4.6 billion years old. Create a Geologic Time Line. Show when different forms of life appeared.

The mantle is 1,800 miles thick. Use the following formula to figure out about how many kilometers that is:

1 mile = 1.61 kilometers

Create a college course of study someone wanting to be a geologist should follow.

Geology Match-up

Match the geological terms on the left to the definitions on the right. Write the letters on the lines that come before the numbers. If you answer correctly, the letters will spell a word that tells what occurs when a dormant volcano "wakes up!"

_____ 1. Lava

_____ 2. Lithosphere

_____ 3. Mantle

_____ 4. Core

_____ 5. Crust

_____ 6. Rocks

_____ 7. Fault

_____ 8. Richter

R. The oceanic and continental crusts and part of the mantle

T. Earth's outer layer

N. Scale used to measure an earthquake's magnitude

E. Magma after it has reached Earth's surface

I. Igneous, sedimentary, and metamorphic are the three main types

U. Layer that comprises most of Earth's mass and volume

O. Fracture in rock formation; one side has moved relative to the other

P. Thought to be composed of iron and nickel

When a dormant volcano becomes active, an _____ occurs.

Minerals

Minerals are inorganic chemical elements or compounds (more than one element) found naturally in the Earth's crust.

Almost all of Earth's materials are made up of minerals. There are about 2,500 different known minerals with new ones being discovered all the time. Each has a specific atomic structure and a characteristic crystal form.

Most minerals are **silicates**. In other words, they are made up of the elements silicon and oxygen, either alone or with other elements. Some silicates are used as construction materials. Examples are glass, concrete, ceramics, brick, and building stones. Other silicates are precious gemstones. Among the most valuable are emeralds, aquamarines, and topaz. Some semi-precious stones, such as agate and jasper, are also silicates.

Create a poster that lists and explains the main physical properties of minerals.

Elements are chemical substances which contain only one kind of atom. Most minerals are made up of more than one element. Those made up of only one element are called native elements. Cite an example of a native element.

Gemstones are crystals of minerals which can be cut and polished into jewels. Create a gemstone word search.

There are about 105 elements, but only 8 elements make up about 99% of Earth's minerals. Two of the 8 make up about 75% of Earth's crust by weight. Which two?

Activities

Some minerals are also metals. Define "metal" and cite three examples.

Unscramble the letters to find out which metal becomes liquid at room temperature.

R E M R C U Y

Draw a picture that shows how this characteristic is put to practical use.

Useful minerals collected for profit are called ores. Create an illustrated poster that shows some minerals we collect and examples of their uses.

In 1812 Austrian geologist Frederick Mohs suggested that geologists use the same minerals as standards for hardness ratings resulting from scratch tests. These tests involve the scratching of one mineral with another to find out how hard that mineral is. Mohs started with the softest mineral and gave it a rating of one. He ended with the hardest, which he rated ten. Although today scientists use much more accurate means of testing in the laboratory, Mohs' scale is still useful for field testing. Research Mohs' scale and create a chart.

AGATE   LEAD   QUARTZ   DIAMOND

Mineral Math-and-Word Clues

Use the math and word clues to figure out some interesting facts about minerals. HINT: You may have to unscramble the letters.

1. This is the only metal that exists as a liquid at room temperature.
   1/2 of HOME   3/4 of CURE   1/2 of VERY   _____

2. The atoms in most minerals form a repeating 3-dimensional pattern causing them to grow into this.
   2/3 of CAR   3/5 of STYLE   1/2 of BALL   _____

3. The overall shape formed by a mineral's crystal is called this. For example, copper has a treelike one.
   3/4 of BITE   1/4 of HOLD   1/5 of ALONE   _____

4. It is the principal ore of aluminum.
   2/3 of BOX   2/5 of UNION   1/2 of PIRATE   _____

5. We use this mineral to make pencils.
   2/5 of ERUPT   1/2 of TIME   2/3 of GATHER   _____

6. This bright yellow nonmetallic mineral forms near hot springs and volcanic vents.
   2/5 of USHER   1/2 of FILE   1/2 of SURE   _____

7. This rare mineral, not discovered until the early 1700s, is even more valuable than gold.
   3/5 of STING   1/2 of SMUG   1/2 of PLAYER   _____

8. This shiny, yellow mineral is sometimes called "fool's gold" because prospectors sometimes mistakenly identify it as gold.
   2/5 of HAPPY   1/2 of TEAR   2/5 of RIGHT   _____

9. The common name for this mineral is salt.
   3/5 of SIGHT   2/5 of ALIEN   1/3 of EAT   _____

10. This very hard mineral is often used in engagement rings.
    3/5 of MONTH   2/3 of DAY   1/2 of DIET   _____

Rocks

Our planet Earth is made up of many different kinds of rocks. What they have in common is that they are composed of minerals. Their different properties are determined by which minerals are present. Geologists classify rocks into three major classes according to how they were formed: igneous, sedimentary, and metamorphic. The subfield of geology that focuses on the study of rocks is called **petrology**.

Igneous rocks form as a result of cooling and solidification of molten matter from the interior of the earth. This matter is called **magma** while it is beneath the surface of the earth and lava once it reaches the surface. Rock that cools and solidifies beneath the surface is called **intrusive rock**. An example of intrusive rock is granite. Rock that forms on the surface of earth as a result of a volcanic eruption is called **extrusive rock**. An example of extrusive rock is basalt.

Sedimentary rocks form as a result of the accumulation and consolidation of sediments. Some sedimentary rocks result because tiny bits of rocks are broken off from larger rocks because of weathering and erosion. These tiny pieces, or sediment, are transported to their place of deposition, where they cement together. An example of this type of sedimentary rock is shale. Other sedimentary rocks are formed as a result of chemical precipitation of minerals from a solution, evaporation of water which contains minerals, and the conversion of organic matter to rock. Limestone is an example of chemically precipitated sedimentary rock. Coal is an example of sedimentary rock formed by the compression of plant remains and layers of sediment. If you see fossils in a rock, it is most likely sedimentary.

Metamorphic rock is rock which has undergone alteration because of exposure to heat and pressure deep within the earth. Any kind of rock—igneous, sedimentary, or metamorphic—can be subject to extreme conditions. The transformation actually makes the rock stronger. Limestone, when exposed to extreme heat and pressure, becomes marble.

BASALT

LIMESTONE

MARBLE

Activities

Examine the roots of the words "igneous," "sedimentary," and "metamorphic." Judge the appropriateness of these terms.

The same type of rock can be changed into more than one type of metamorphic rock depending on the amount of heat and pressure it is exposed to. Create a chart showing the kinds of rock shale can become.

Coal begins with the remains of ancient swamp plants. Chart the transformation of these plants from peat to anthracite.

Create a travel brochure for a "Rocky Tour of the World." Include monuments made of igneous, sedimentary, and metamorphic rocks.

Draw a picture of a famous European natural site made of seashell remains.

Draw a picture of a conglomerate rock.

Gradual Changes in the Earth's Surface

The surface of the Earth is always changing. Most of these changes take place over long periods of time. These gradual changes are caused by weathering, erosion, and sedimentation.

WEATHERING

Weathering breaks rocks into smaller pieces. There are two types of weathering—mechanical and chemical. Running water, temperature changes, wind, near-ocean shore waves, glacial ice, gravity, and the activities of plants and animals are all agents of mechanical weathering. Especially important is the alternating of freezing and thawing in porous rocks, which allow water to seep in. When the water freezes, it expands. This can weaken the rock, causing pieces to break away. This is called frost wedging. Sometimes chemicals found in the air and water dissolve and wear away rock. Unlike mechanical weathering, chemical weathering changes the rocks.

EROSION AND SEDIMENTATION

Erosion and sedimentation complement each other. **Erosion** is the general term for the processes that wear away rock materials and remove them from the surface of the earth. **Sedimentation** is the process of taking those materials and depositing them somewhere else. Before the material is deposited, the material often travels great distances.

Erosion is caused by weathering, chemical action, and abrasion. When it occurs on or near the surface of the earth because of exposure to the atmosphere, it is call **subaerial erosion**. If it results from the actions of water currents on the bottoms of standing water, it is referred to as **subaqueous erosion**.

At some point, of course, the transporting agent—wave, wind, gravity, current, etc.—is forced to deposit its sediment. For example, if material is being carried down a hill by gravity, at some point it reaches the bottom and can go no further. Running water may deposit its sediment when currents slacken. Evaporation may result in deposition.

Weathering is dependent upon several factors: climate, topography, vegetation or lack of it, and the composition and texture of the rock itself. For example, the climate of Bryce Canyon in southern Utah is dry. When it does rain, however, the running water quickly forms gullies because the rock is soft and the vegetation is sparse. Choose a location and explain how one or more of the mentioned factors affected the formation of the significant feature.

Activities

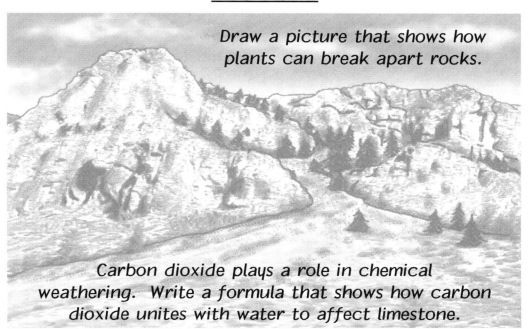

Draw a picture that shows how plants can break apart rocks.

Carbon dioxide plays a role in chemical weathering. Write a formula that shows how carbon dioxide unites with water to affect limestone.

Draw a picture that illustrates the result of oxygen and moisture on iron-bearing sandstone.

Create a detailed travel poster for a tour of the U.S.A. including at least 5 interesting land formations caused by erosion or weathering.

Analyze the following metaphor: "Waves and currents are the ocean's teeth."

Glaciers and Glaciation

A **glacier** is a large mass of permanent ice. It forms on land because of the compacting and recrystallization of snowflakes. Gravity causes the snow to move down a slope. Its own thickness causes the accumulated mass to spread outward in all directions. Most glaciers are found in high altitudes and latitudes. Australia is the only continent to have no glaciers.

There are four main types of glaciers. **Valley glaciers**, also called mountain or alpine glaciers, flow down a valley. The moving ice from the mountain snowfields follows the valleys once cut out of the mountains by the running water of streams. Glaciers of this type are much longer than they are wide. They vary greatly in length—from about half a mile to well over 50 miles in length! Valley glaciers are very plentiful. There are over 1,200 found in the Alps alone.

Piedmont glaciers are found only in very high latitudes. They are formed when a valley glacier becomes so large that it leaves the valley. Sometimes two or more valley glaciers flow together and form quite large piedmont glaciers. A well-known piedmont glacier is Malaspina Glacier in Alaska. It covers an area of about 1,600 square miles and is about 2,000 feet thick! However, most piedmont glaciers have disappeared.

Ice sheets are dome-shaped glaciers with a central area of accumulation. They move outward in all directions from that area. Small ice sheets are called **ice caps**. Huge ones—those over 50,000 square kilometers (19,300 square miles)— are called **continental glaciers**.

Glaciers change the topography of the land through erosion, transportation, and deposition. Glacial deposits are known as **drift.** As valley glaciers remove more and more rock from the top of the mountains, large, bowl-like depressions are formed. These depressions are called **cirques.** Glaciers also change V-shaped valleys into U-shaped ones by grinding away at the bases and leveling the valley floor. Sometimes tributary valleys are left hanging; when these valleys contain streams, **waterfalls** are formed.

Activities

Locate Malaspina Glacier on a map.
Figure out how many square
kilometers it covers and how
many meters thick it is.
Use the following formulas:

1 square mile = 2.6 square kilometers
1 foot = .30 meters

Locate the
world's highest
waterfall on
a map.

Only two continental glaciers are in existence today.
Find out where they are located. Locate them
on a world map.

Unscramble the
letters to find out
what we call a massive body of ice
that has broken away from a glacier.

GUCIEBR

Draw a map showing the
extent of ice coverage during
the Great Ice Age.

Plains and Plateaus

Along with mountains, plains and plateaus are the most common types of landforms on Earth. Research both. Create a Venn diagram in which you compare and contrast the two landforms.

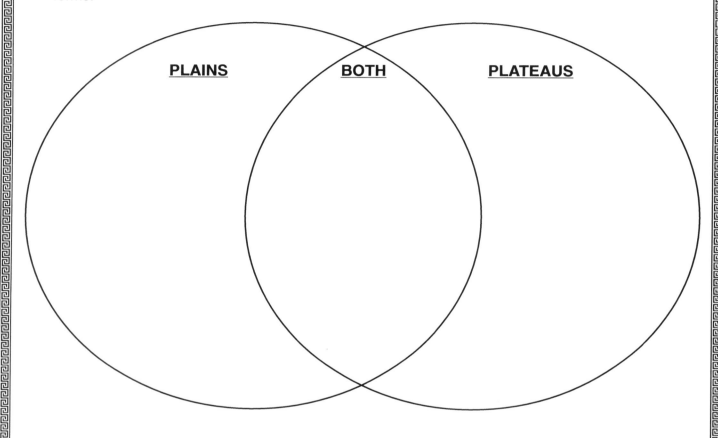

PLAINS   **BOTH**   **PLATEAUS**

Define mesas, buttes, and rock spires.

Draw a series of pictures that shows what happens when erosion keeps wearing away a plateau.

Mountain Building

A mountain is a large mass of rock which has been moved higher than its original position. It has steep sides and a relatively narrow summit area. Some geologists limit the term "mountain" to those landforms which reach a height of at least 2,000 feet above the surrounding area.

Geologists classify mountains according to the way in which they were formed. All are the result of rock having changed position in some way. **Erosional mountains** are the result of the wearing away of less resistant rocks of a plateau. **Volcanic mountains** are built up by lava as cones or shields. **Dome mountains** are formed when intrusive rock forms deep within the earth and pushes the crust upward. **Fault-block mountains** result from the vertical movement, or faulting, of large blocks of the earth's crust, lifting the blocks to high levels. **Folded mountains** are formed because of the interaction of the borders of the moving plates that make up earth's crust.

Activities

Mountains have a great impact on the climate, population, and economy of a region. Research and report on the impact of mountains.

Find examples of different types of mountains and create a poster illustrating the different structural classifications.

Changes in the earth's crust which lead to mountain building occur in two main ways: volcanism and diastrophism. Explain each.

Create a chart that shows the relationships among the following: isolated mountain, range, system, chain, and cordillera.

Volcanoes

A volcano is a **vent,** or opening, in the earth's crust from which molten rock and gas erupt. While beneath the surface of the earth, this mixture of molten rock and the gases it carries is called **magma.** The magma which erupts from a volcano is formed in the upper layer of the mantle and the lower layer of the crust, where rocks are exposed to a great deal of heat and pressure. The pressure pushes the rocks toward the surface of the earth. These moving rocks become so hot that some of the material melts, forming magma.

If enough rock melts—in other words, if enough magma forms—the magma rises to the surface of the earth, where it finds its way through the vents. Some also flows through cracks in the earth, called **fissures.** The molten rock and gases, called **lava** once it reaches the surface of the earth, is discharged through the vent. A volcanic eruption has occurred!

Each time a volcano erupts, the lava builds up around the vent. This is how the volcano grows, or builds itself up. There are different types of volcanoes depending on how the lava flows. While some volcanoes erupt violently, others emit quiet flows of lava. The way they erupt depends on the kind of magma. There are two basic kinds: granite and basalt.

Granite magma, although light, is viscous, or sticky, and does not flow easily; therefore, it holds a lot of gases and steam. These gases often remain trapped underground until enough gas pressure builds up to force their way out of the vent. This causes an explosive type of eruption. Because the lava flows slowly, it often cools and hardens before leaving the vent and closes up the opening. A volcano formed from loose rocks thrown from a volcanic vent is called a **cinder cone**. The rocks fall back into the bowl-shaped depression that forms at the top, called a **crater**, and build up the steep-sided cone.

Basalt magma, although heavier, flows more easily. It allows more steam and gas to escape while underground. The gas pressure does not build up as it does with granite magma, so there is no explosion. When it reaches the surface, the lava flows quickly and quietly away from the vent before cooling and hardening. **Shield volcanoes**, which have broad, gently sloping cones with flattened tops, result. It takes many layers of hardened lava flows to make a whole mountain. Most shield volcanoes rise from the ocean bottom.

Some volcanoes are composites, or mixtures, of explosive eruptions and free flows of lava. Volcanoes built up in this way are called **stratovolcanoes,** or **composite cones.** Composite cones are the most common.

Not all volcanoes are active. Some are believed to be **extinct** because they have not erupted for such a long time. But a volcano that seems to be extinct might only be **dormant,** or at rest. If the underground pressure builds up enough, the volcano might erupt!

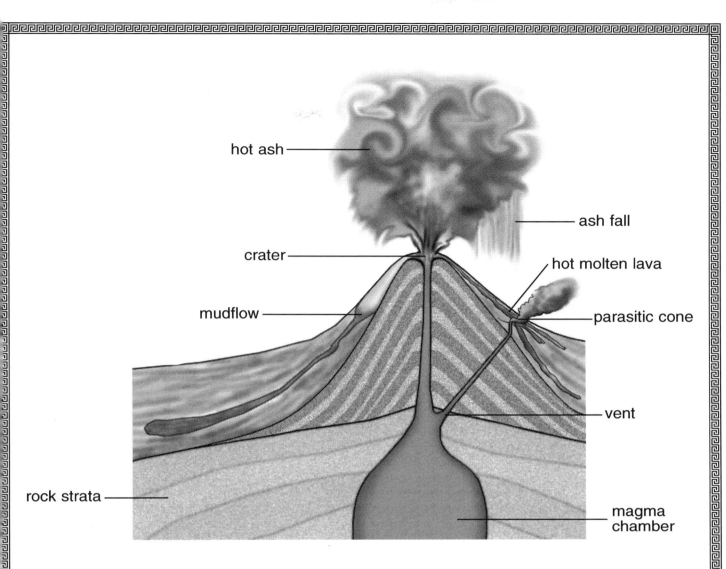

hot ash

ash fall

crater

hot molten lava

mudflow

parasitic cone

vent

rock strata

magma chamber

Activities

Mount St. Helens
in the state of Washington
erupted in 1980. Most of the people
evacuated the area; nevertheless, 35 people
were killed and another 25 were missing.
Among the victims of the volcano was
84-year-old Harry S. Truman, who refused to
leave his home of 50 years. Suppose you had
a chance to talk to Mr. Truman about his
choices. What advice would you have given him?

Activities

Create a flipbook of blown-out pieces of lava, which are called pyroclastics.

Draw a poster that illustrates the following types of volcanic cones: composite cones, shield cones, and cinder cones. Cite an example for each.

Mauna Loa is Earth's largest volcano. It is even taller than Mount Everest. Draw a picture of Mauna Loa. Label its height above sea level and also its height above the ocean floor. Locate Mauna Loa on a map.

Use Crater Lake to explain what a caldera is. Sketch a picture.

The principal area of volcanic activities encircles the Pacific Ocean. Locate the Pacific Rim of Fire on a map.

Lava erupts at temperatures ranging from 850 to 1,250 degrees Celsius. Convert these temperatures to degrees Fahrenheit using the following formula:

Fahrenheit = Celsius x 9/5 + 32

Earthquakes

An **earthquake** is a trembling or shaking of the earth's surface caused by the fracture of rock within the earth. Rocks in the earth's crust are under a lot of pressure and gradually bend. If the pressure builds up enough, the rocks break, or fracture, and then snap back. This breaking and snapping back is called an **elastic rebound.** It is the elastic rebound which causes the sudden **tremor** within the earth that we call an earthquake. Most earthquakes occur along cracks, or **faults,** in the crust. Rocks along these faults shift and break more easily. Many of these faults occur along faults located beneath the sea, especially near the coasts of the continents or the island chains.

Scientists explain most earthquakes by the **tectonic plate theory.** "Tectonic" comes from a Greek word which means "structure." According to the theory there are about six huge plates and several smaller ones. These plates, which make up the lithosphere, move extremely slowly over the asthenosphere. Each has a leading edge and a trailing edge. When the leading edge of one plate goes under the trailing edge of another, faults develop in the trailing edge.

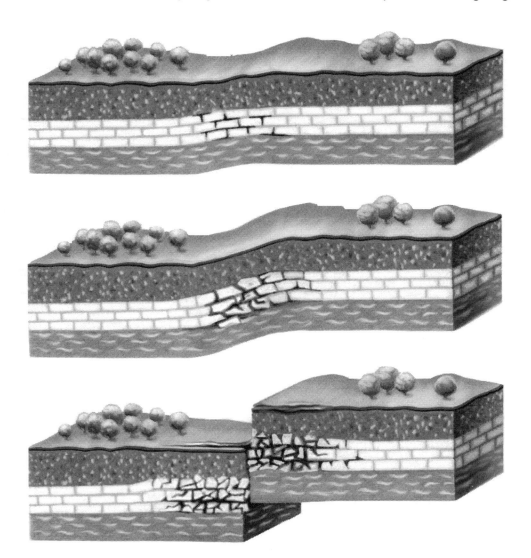

Activities

A *seismologist is a scientist who studies earthquakes and related phenomena. See how many words of three or more letters you can form by using the letters in "seismologist." Do not use the "s" to form plurals or to change the verb forms.*

S E I S M O L O G I S T

Geologists learn a lot about earth's structure from the study of the waves generated by earthwaves, called seismic waves. There are three kinds of seismic waves: P waves, S waves, and L waves. Make a chart explaining these.

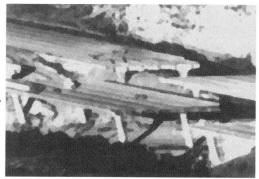

Explain what is meant by epicenter.

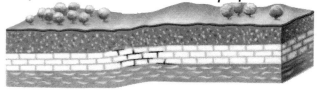

The Mercalli scale measures the intensity of earthquakes by assessing the damage. It ranges from I to XII. Create an illustrated poster of the Mercalli scale.

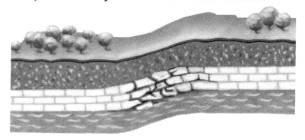

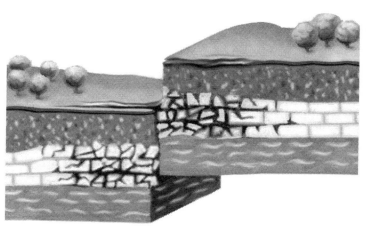

In 1935 Charles F. Richter devised a scale to describe the intensity of an earthquake. It measures the movement of the ground from 0 to 9. Explain why an earthquake that measures 4 is not twice as great as one that measures 2.

Shaken-up Vocabulary

The following terms about earthquakes were "shaken up." Use the clues to help you unscramble the letters and figure out the words.

1. It describes something relating to or caused by an earthquake.

   I E S M I C
      __ __ __ __ __ __ __
              4

2. It's a crack or break in rock in which one side has moved relative to the other.

   L F A U T
      __ __ __ __ __
             7

3. This scale measures an earthquake's magnitude.

   C I R H E T R
      __ __ __ __ __ __ __
         3

4. Vibrations caused by earthquakes are in this form.

   V A W S E
      __ __ __ __ __
           2

5. The Mercalli scale measures the intensity of earthquakes by rating this.

   E A D M A G
      __ __ __ __ __ __
          5   9

6. Undersea earthquakes can cause this.

   U S T N M A I
      __ __ __ __ __ __ __
        1

7. The study of earthquakes is a subfield of this branch of science.

   O E G O L Y G
      __ __ __ __ __ __ __
           6   8   10

Now write the letters above the numbers and fill them in the appropriate spaces below. The letters will spell out the subfield involved in the study of earthquakes and Earth's interior.

__ __ __ __ __ __ __ __ __ __
1   2   3   4   5   6   7   8   9   10

Other Geologic Phenomena

Activities

Fumaroles are often found in volcanic areas. Explain what a fumarole is.

Define the terms "hot spring" and "geyser." Identify the most famous geyser in the U.S.A. and judge its name.

Tsunamis are sometimes called tidal waves. Evaluate the use of this term.

Use the change that occurs in tsunamis from the time they are in open ocean to when they approach land to explain why they are so damaging.

Draw a picture that shows what occurs when part of a hillslope weakens and is unable to support its own weight.

Unscramble the letters to learn what it is called when a large amount of snow and ice as well as rock and soil fall suddenly and rapidly down a mountain.

H V A A L A N C E

Plate Tectonics

Plate tectonics is a rather new concept related to the theory of continental drift. Around 1912 meteorologist and geophysicist Alfred Wegener first presented his theory that the continents were once a huge supercontinent and that they slowly drifted apart. His theory was not seriously considered until the late 1950s.

Today most scientists accept the theory of plate tectonics. They believe that the lithosphere is divided into about six huge, rather rigid plates and up to twelve smaller ones. Most plates contain a continent and adjacent portions of the ocean floor. They are thought to be about 44 to 50 miles (7 to 80 kilometers) thick under the oceans and about 62 to 93 miles (100 to 150 kilometers) thick under the continents. The plates float on a viscous underlayer in the mantle, called the asthenosphere. A lot of seismic activity and volcanism occur at their boundaries. The movement of the plates also cause the continents to drift and the continents and ocean basins to change their size and shape.

There are several types of boundaries between plates: spreading boundaries, or mid-oceanic ridges; convergent boundaries; and shear boundaries, or transform faults. **Mid-oceanic ridges** are *constructive.* It is here where seafloor, or ocean-floor, spreading takes place. As the plates move apart, partially melted mantle rock wells up and fills in part of the rift, resulting in the creation of new crust. **Convergent boundaries,** on the other hand, are *destructive.* In places known as subduction zones, the edge of one plate descends beneath another. The plate is consumed as it plunges into the hot magma, forming deep, narrow undersea valleys, known as ocean trenches. Convergent plate boundaries also include zones of collision. Colliding may cause earthquakes, volcanoes, and folding. **Transform faults,** or shear boundaries, are *neither constructive nor destructive.* They occur when the plates slide past each other, causing neither the creation of new lithosphere nor the destruction of existing lithosphere.

Activities

Draw a map of the earth's tectonic plates.

Explain what is meant by Pangaea.

Define "rift valley." Locate the Great Rift Valley of Africa.

Draw a picture that shows what is meant by "continental shelves." Explain how they are exploited commercially.

Geology Fact vs. Opinion

A fact is a statement which can be proved. An opinion is a belief not substantiated by proof. Mark each statement with an "F" for fact or an "O" for opinion.

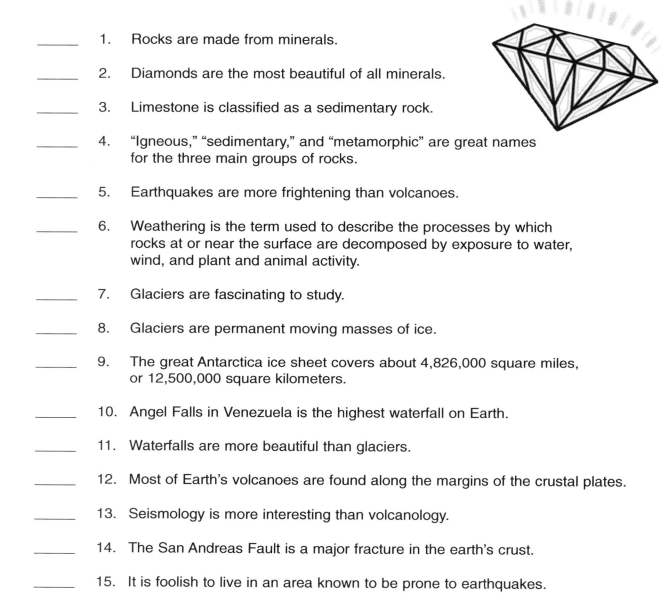

_____ 1. Rocks are made from minerals.

_____ 2. Diamonds are the most beautiful of all minerals.

_____ 3. Limestone is classified as a sedimentary rock.

_____ 4. "Igneous," "sedimentary," and "metamorphic" are great names for the three main groups of rocks.

_____ 5. Earthquakes are more frightening than volcanoes.

_____ 6. Weathering is the term used to describe the processes by which rocks at or near the surface are decomposed by exposure to water, wind, and plant and animal activity.

_____ 7. Glaciers are fascinating to study.

_____ 8. Glaciers are permanent moving masses of ice.

_____ 9. The great Antarctica ice sheet covers about 4,826,000 square miles, or 12,500,000 square kilometers.

_____ 10. Angel Falls in Venezuela is the highest waterfall on Earth.

_____ 11. Waterfalls are more beautiful than glaciers.

_____ 12. Most of Earth's volcanoes are found along the margins of the crustal plates.

_____ 13. Seismology is more interesting than volcanology.

_____ 14. The San Andreas Fault is a major fracture in the earth's crust.

_____ 15. It is foolish to live in an area known to be prone to earthquakes.

Geologic Syllogisms

A syllogism is a form of deductive reasoning. There are a major premise (A), a minor premise (B), and a conclusion (C). For an argument to be valid, the conclusion, C, must be based on and consistent with statements A and B. Note: In this exercise do not be concerned with whether the facts are true or false. Mark each argument as either Valid (V) or Invalid (I).

1. A. All igneous rocks were formed by cooled and hardened magma.
   B. Granite is igneous rock.
   C. Therefore, granite was formed by cooled and hardened magma.

   _____

2. A. Some minerals are metals.
   B. Diamonds are minerals.
   C. Therefore, diamonds are metals.

   _____

3. A. All rocks which have been transformed from a pre-existing rock by heat and/or pressure are called metamorphic rocks.
   B. Shale, when exposed to extreme pressure, transforms into slate.
   C. Therefore, slate is a metamorphic rock.

   _____

4. A. Some volcanoes are extinct.
   B. Mt. St. Helens is a volcano.
   C. Therefore, Mt. St. Helens is extinct.

   _____

5. A. Many geologists study chemistry.
   B. Zachary studies chemistry.
   C. Therefore, Zachary is a geologist.

   _____

6. A. Earthquakes are geologic hazards.
   B. Volcanoes are geologic hazards.
   C. Therefore, earthquakes are volcanoes.

   _____

7. A. All rocks are made up of minerals.
   B. Granite is rock.
   C. Therefore, granite is made up of minerals.

   _____

An Ancient Myth

Many ancient peoples believed that geological phenomena were caused by supernatural beings. Decipher the code and read the passage describing an ancient Roman myth.

A =	E =	I =	M =	Q =	U =	Y =
B =	F =	J =	N =	R =	V =	Z =
C =	G =	K =	O =	S =	W =	
D =	H =	L =	P =	T =	X =	

CP   CPEKGPV   OAVJ

KP VJG OGFKVGTTCPGCP UGC KU C ITQWR QH
XQNECPKE OQWPVCKPU PCOGF VJG NKRCTK KUNCPFU.
VJG OQUV UQWVJGTP QH VJGUG OQWPVCKPU KU
ECNNGF XWNECPQ. XWNECP YCU CNUQ VJG PCOG
QH VJG CPEKGPV TQOCPU' IQF QH HKTG.

GXGPVWCNNA, VJG TQOCPU DGICP VQ
KFGPVKHA XWNECP YKVJ VJG ITGGM IQF
JGRJCGUVWU. JGRJCGUVWU YCU IQF QH VJG HQTIG.
JGRJCGUVWU, QT XWNECP, YCU QHVGP RQTVTCAGF CU
C DNCEMUOKVJ, NCDQTKPI QXGT C JQV HKTG CPF
HCUJKQPKPI DGCWVKHWN VQQNU CPF QTPCOGPVU.
YJGP HKTG ECOG QWV QH VJGKT OQWPVCKPU, VJG
CPEKGPV TQOCPU VJQWIJV KV YCU URCTMU HTQO
XWNECP'U HWTPCEG. VQFCA YG UVKNN ECNN VJGUG
OQWPVCKPU "XQNECPQGU" CHVGT VJG CPEKGPV IQF
QH HKTG.

Geology Word Search

Create a geology word search using at least twenty words from the list below. Add other geology terms if you wish. Exchange with classmates to solve.

avalanche	fault	limestone	quartz
basalt	glacier	magma	rock
crust	granite	mantle	sedimentary
crystal	hot spring	metamorphic	seismic
earthquake	iceberg	mineral	tsunami
epicenter	igneous	mountain	volcano
erosion	lava	plates	weathering

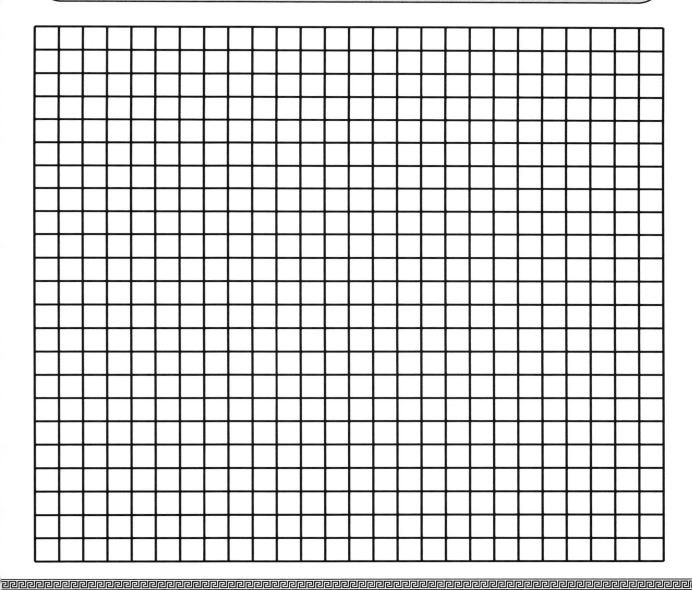

Geology Scrambled Words

Unscramble the letters to figure out these geology terms.

1. H T A E R U A Q E K _____

2. C O V L N A O _____

3. N G I O E S U _____

4. U R E P I T O N _____

5. A L V A _____

6. G A M M A _____

7. R E O S I N O _____

8. A M N L T E _____

9. R C S U T _____

10. N I A T N U O M _____

11. V A A L A C N H E _____

12. N A L D S I L E D _____

13. E G Y E S R _____

14. E G O O L I G S T _____

Odds 'n Ends

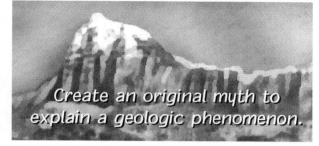
Create an original myth to explain a geologic phenomenon.

Make a model volcano out of papier-mâché.

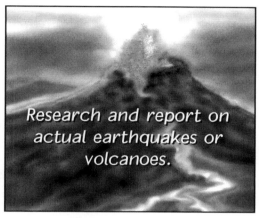
Research and report on actual earthquakes or volcanoes.

Locate on a map the site of three active volcanoes.

The principal area of volcanic activities encircles the Pacific Ocean. Draw a map of the Pacific Rim of Fire.

Start a rock collection. Find at least 5 different rocks and identify them. Use a rock-and-mineral guide to help you.

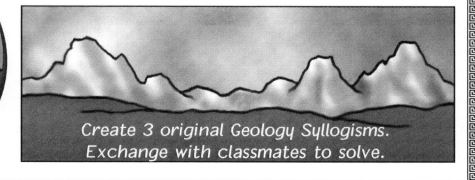
Create 3 original Geology Syllogisms. Exchange with classmates to solve.

Write a sentence in which you personify an earthquake or a volcano.

An acrostic is a non–rhyming poem in which the first letter of each line spells a name, object, place, or idea. Create an acrostic about one of the following: earthquake, volcano, mountain, avalanche, tsunami, or erosion.

Create an illustrated Geology ABC Booklet.

Write an alliterative sentence about rocks and/or minerals. Your subject may be specific or general.

Create three geology riddles or jokes. Exchange with classmates to solve.

Research and report on an interesting geologic landmark or event. Some possibilities are the Grand Canyon, Old Faithful, the eruption of Mt. Vesuvius, and the Badlands.

Our Ever-Changing Earth
Crossword Puzzle

ACROSS
5. The crust is made up mostly of silicate ___.
7. The central portion of the earth.
8. A large ocean wave caused by an underwater earthquake or volcanic eruption.
9. One of Earth's principal land masses.
10. Rock formed by solidification from a molten state.
12. The thin, exterior portion of the earth.
13. This layer comprises the bulk of Earth's mass and volume.
14. A vent in the earth from which molten rock and gas erupt.
19. A rock which has been changed by extreme heat and pressure.
20. A bend in a layer of rock.
22. Molten rock at the earth's surface.
23. A trembling or shaking movement of the earth's surface.
24. Caused by or otherwise having to do with an earthquake.
26. The downward slide of a mass of earth & rock.

DOWN
1. The universal continent.
2. Wears away rocks, removing them from one area of the earth's surface.
3. Theory that the outermost shell of Earth is divided into rigid plates.
4. The study of the planet Earth.
5. Landforms with steep sides and a narrow summit area.
6. Chemical and physical processes by which rocks exposed to the weather undergo changes and break down.
11. Rocks caused by the accumulation and consolidation of layers of loose sediment.
15. A slide of a large mass of snow and ice.
16. Type of valley formed when land sinks between two fairly parallel faults.
17. San Andreas, for example.
18. A huge mass of moving ice.
21. Molten rock under the earth's crust.
25. Powdery matter ejected by a volcanic eruption.

What's the Question?

What's the Question? is similar to the TV show "Jeopardy" in that the information is given in the form of the statement and the student responses are in the form of questions. The questions in Part I are worth 5 points each, and those in Part II are worth 10 points each.

Divide the class into teams of 4 or 5 students. The teacher may act as leader, or you may want to choose a student leader. The leader asks the first group a question from Part I. Whoever raises his or her hand first gets to answer. If the student answers correctly, 5 points are added to the team total. If the student answers incorrectly, 5 points are deducted. If no one wants to answer, the leader gives the correct answer and the total remains the same. If a team does not give a correct answer, the same question is then asked to the next group. If no group gets it right, the leader gives the correct answer.

When all the questions from Part I have been completed, the same rules are followed for Part II.

ANSWERS TO "THE HUMAN BODY WHAT'S THE QUESTION?"

PART I: 5 points each
1. What is the crust?
2. What is the mantle?
3. What is the lithosphere?
4. What is the core?
5. What is the mantle?
6. What is petrology?
7. What is seismology?
8. What is volcanology?
9. What is mineralogy?
10. What are physical & historical geology?
11. What are silicates?
12. What are metals?
13. What are gemstones?
14. What is diamond?
15. What is a crystal?
16. What are minerals?
17. What is igneous?
18. What is sedimentary?
19. What is metamorphic?
20. What is limestone?
21. What is an earthquake?
22. What is a volcano?
23. What is a landslide?
24. What is an avalanche?
25. What are tsunamis?

PART II: 10 points each
1. What are erosional mountains?
2. What are volcanic mountains?
3. What are dome mountains?
4. What are fault-block mountains?
5. What are folded mountains?
6. What is magma?
7. What is lava?
8. What is a cinder cone?
9. What are shield volcanoes?
10. What are stratovolcanoes?
11. What is a fault?
12. What is the epicenter?
13. What is the Richter scale?
14. What is damage done by an earthquake?
15. What are seismic waves?
16. What is the lithosphere?
17. What is the asthenosphere?
18. What is Pangaea?
19. What are mid-oceanic ridges?
20. What is seafloor (ocean-floor) spreading?
21. What is the Mohorovičić discontinuity? (Moho)
22. What is weathering?
23. What is erosion?
24. What is a fault?
25. What is continental drift?

What's the Question?

Earth's Structure & Composition	"–Ologies"	Minerals	Rocks	Geological Hazards
1. It's the outer layer of the earth.	6. It's the study of rocks.	11. This type of compound occurs in most rocks except limestone and dolomite.	16. Rocks are made of them.	21. It's a sudden violent vibration in Earth's crust.
2. It's the layer between the crust and the core.	7. It's the study of earthquakes.	12. Shiny, moldable elements that will conduct electricity are called this.	17. This main type of rock forms when magma cools and hardens.	22. It's an opening through which lava erupts and also the mountain formed by the eruption.
3. This rigid outer layer is made up of the crust and the upper mantle.	8. It's the study of volcanoes.	13. It's a general term for stones that are used as jewels when polished and cut.	18. This main type of rock forms from pieces of rock and other deposition of sediment.	23. It's a massive downward and outward movement of slope-forming materials.
4. The center of the Earth consists of a solid inner one and a molten outer one.	9. It's the study of the classification of minerals.	14. It's the hardest natural substance on Earth.	19. This main type of rock was transformed from pre-existing rock by heat or pressure.	24. It's a rapid downward movement of ice and snow.
5. This layer comprises most of Earth's mass and volume.	10. They're the two main divisions of geology.	15. Minerals form regular, flat-sided shapes called this.	20. Marble was once this sedimentary rock.	25. These catastrophic ocean waves are generated by underwater earthquakes or volcanoes.

What's the Question?

Part II: 10 points

Mountain Types	Volcanoes	Earthquakes	Plate Tectonics	Potpourri
1. They result from the wearing away of less resistant rocks of a plateau.	6. Molten rock is called this when beneath the Earth's surface.	11. It's a fracture in a once-continuous rock formation where adjacent surfaces are displaced relative to each other.	16. This rigid layer of crust and part of the mantle is divided into plates.	21. It's the boundary between the crust and the underlying mantle.
2. They are built up of lava.	7. Molten rock is called this once it reaches the surface.	12. It lies directly above the origin of an earthquake.	17. The tectonic plates float on this lower, soft layer of the mantle.	22. It's the breakdown of rocks from mechanical and chemical processes of the weather, such as the freezing and thawing of ice.
3. They form when intrusive rock pushes the crust upward.	8. This type of volcano is formed from loose rocks thrown from a volcanic vent.	13. The logarithmic scale measures the intensity of an earthquake.	18. It's what we call the supercontinent of about 275 to 175 million years ago.	23. It's the gradual wearing away of rock by weathering, corrosion, and abrasion.
4. They result from the vertical movement of large blocks of crust.	9. Basalt magma forms these volcanoes with broad, gently sloping cones.	14. The Mercalli scale assesses this.	19. New oceanic lithosphere is formed along these undersea mountain ranges when plates move apart.	24. San Andreas is a famous one.
5. They result from the interaction of borders of the crust's moving plates.	10. They are also called composite cones.	15. "P," "S," and "L" are used to describe these.	20. Term given to the process of forming a new ocean-floor layer at a mid-oceanic ridge.	25. This theory about the movement of continents is closely related to the theory of plate tectonics.

Answers and Background Information

"-Ologies" (Page 8)
1. O   2. C   3. E   4. A   5. N   6. O   7. G   8. R   9. A   10. P   11. H   12. Y

Earth's Structure and Composition (Pages 9–11)

The layers of the atmosphere from lowest to highest are troposphere, stratosphere, mesosphere, thermosphere, and the ionosphere. Beyond that is the exosphere, which reaches up to about 18,000 miles.

The Moho is a boundary layer between the crust and underlying mantle. This layer is marked by a sharp alteration in the velocity of earthquake waves passing through that region.

The mantle is about 2,898 kilometers thick.

Geology Match-up (Page 12)
1. E   2. R   3. U   4. P   5. T   6. I   7. O   8. N

Minerals (Pages 13–14)

Answers will vary, but some of the physical properties of minerals which should be mentioned are hardness (resistance to scratching); luster (appearance when light is reflected from its surface); specific gravity (relative weight compared with an equal volume of water); color; cleavage (tending to split along certain planes); and fracture (the way it breaks other than cleavage). Other properties include transparency, streak, striations, tenacity, fusibility, heat conductivity, taste, odor, feel, and magnetic and electrical properties.

Gold, silver, copper, and diamonds are native elements.

The eight elements that make up 99% of Earth's minerals are oxygen, silicon, aluminum, iron, magnesium, calcium, potassium, and sodium. Oxygen and silicon make up 75% of Earth's crust by weight.

Metals are shiny, moldable elements that will conduct electricity. Gold, silver, copper, nickel, iron, and platinum are a few metallic elements.

Mercury exists in a liquid state at room temperature. It is used in thermometers.

Mohs' scale lists the minerals in the following order from softest to hardest: talc, gypsum, calcite, flourite, apatite, orthoclase, quartz, topaz, corundum, and diamond.

Mineral Math and Word Clues (Page 15)
1. mercury	3. habit	5. graphite	7. platinum	9. halite
2. crystal	4. bauxite	6. sulfur	8. pyrite	10. diamond

Rocks (Page 17)

The root "-ign" is derived from the Latin word *ignis,* meaning "fire." "Sedimentary" comes from the Latin words *sedimentum,* meaning "act of settling," and the verb *sedere,* which means "to settle." "Metamorphic" comes from the Greek words *meta,* meaning "change," and *morphe,* meaning "form."

If shale is exposed to moderate heat and pressure, it can be transformed into slate. If exposed to high heat and pressure, it can become schist. If exposed to extreme heat and pressure, it can become gneiss.

When plants decay in the mud, they turn into a partially carbonized material resembling moist tobacco, called peat. Sedimentary rocks forming on the peat may crush it and turn it into lignite. If more rocks apply pressure, the lignite becomes bituminous coal. Still more pressure turns it into anthracite. Anthracite is very hard and shiny. The harder the coal, the more energy released when burned.

Brochures will vary, but an example of each rock type follows: The Great Wall of China was made in part with granite, an igneous rock. The Egyptian and Mayan pyramids were built with limestone, a sedimentary rock. The Taj Mahal was built in marble, a metamorphic rock.

The White Cliffs of Dover are a famous limestone formation on the coast of England. This type of limestone was formed by seashell remains cemented together.

A conglomerate is a rock consisting of pebbles and gravel embedded in a loosely cementing material.

Gradual Changes in Earth's Surface (Page 19)

Plants can grow through cracks in rocks, breaking apart the rocks.

Carbon dioxide plus water make carbonic acid: (H_2CO_3). Carbonic acid can decompose limestone.

The affect of oxygen and moisture on iron-bearing sandstone is rust. When the rocks rust, they fall apart.

Glaciers and Glaciation (Page 21)

Malaspina Glacier covers about 4,160 square kilometers and is about 600 meters thick.

Angel Falls in Venezuela is the world's highest waterfall.

The continental glaciers are the ice sheets of Greenland and Antarctica.

A massive body of ice that has broken away from a glacier is called an iceberg.

Ice Coverage During the Great Ice Age

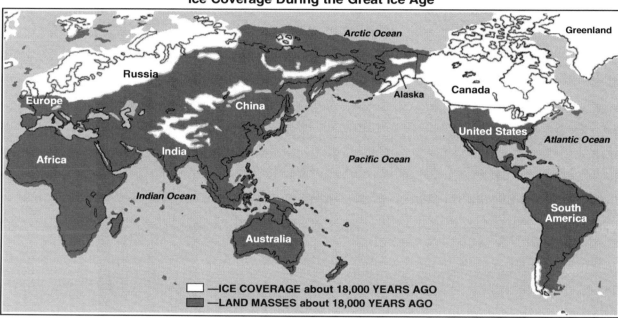

—ICE COVERAGE about 18,000 YEARS AGO
—LAND MASSES about 18,000 YEARS AGO

Plains and Plateaus (Page 22)

PLAINS

1. Surface not deeply cut by canyons or river valleys
2. Top and bottom are only a few hundred feet apart
3. May originate from ocean floor, from a lake basin, or from glacial and river activities
4. Drainage areas may be broad, or spread out

BOTH

1. Large
2. Fairly flat
3. Rock layers beneath the land-forms are flat
4. Formed where flat layers of rocks were deposited
5. The rocks on which deposited are either volcanic or sedimentary

PLATEAUS

1. Small mountains and canyons are present
2. Top of mountains and bottom of canyons are several thousand feet apart
3. May be formed from volcanism
4. May have mountain ranges, deep gorges, and other landforms
5. Erosion main cause of relief
6. Drainage areas not spread out
7. Streams cut into the rock

Mesas are the flat-topped elevations of plateaus. When mesas become reduced because of erosion, they become buttes. If buttes are reduced to a small enough size, they become rock spires, which eventually break and collapse.

If enough erosion occurs, the plateau will become a plain.

Mountain Building (Page 23)

The following might be included in a report on the impact of mountains: Mountains intercept prevailing winds, causing precipitation on the windward side. They are not suitable for agriculture. The colder climate and rarified atmosphere tend to lead to lower population. They are often rich in minerals. Mountains act as natural borders. They make travel more difficult.

Volcanism is the movement of liquid rock, called magma, from the interior of the earth towards the surface. Diastrophism is the movement of solid rocks in the earth's crust, causing the rocks to bend or break.

An isolated mountain is not associated with any others in age, origin, or geographic alignment. A mountain range is a long, narrow ridge of mountains of similar age and origin. (If not in a belt, it is called a group.) A mountain system comprises ranges and groups of similar age and origin. A mountain chain comprises an elongated group of systems. A complex of ranges, groups, and systems with differing ages and origins is called a cordillera.

Volcanoes (Page 26)

The following should be included in the flipbook: volcanic bombs, scoria, cinders, pumice, lapilli, and volcanic dust and ash. The smallest bits are volcanic dust and ash. The frothy masses, which are so light they float, are called pumice. Lapilli are small, hard stones. Cinders and large chunks, called scoria, are burnt, crustlike lava with large holes from the escaping gases.

Mauna Loa is located in the Hawaiian Islands. It projects 13,677 feet (4,170 meters) above sea level and over 29,000 feet (8,850 meters) above the ocean floor.

Crater Lake is an example of a caldera, which forms when the top of a volcano collapses after an eruption.

The range of temperature for erupting lava is 1,562 to 2,282 degrees Fahrenheit.

Earthquakes (Page 28)

The following are among the words that can be formed by using the letters in "seismologist": emit, gem, get, gist, gloom, gloss, let, lime, list, lit, log, logo, loom, loose, loot, lose, loss, lost, melt, met, mile, mist, mite, mole, molt, moo, moose, moot, moss, most, motel, ogle, oil, oleo, osmosis, set, silt, sit, site, slime, smelt, smile, smite, smote, soil, sole, solo, some, soot, stem, stole, stool, tile, time, toe, tog, toil, tool, and toss.

P, or primary, waves, are also called compression, or push, waves. They travel the fastest and, therefore, are the first to be recorded on a seismogram. **S, or secondary, waves,** are also called transverse, or shear, waves. S waves cause the earth to vibrate perpendicularly to the direction of their motion. **L, or long, waves,** are also called surface waves. Generated from P and S waves, L waves are the most destructive. (Surface waves are sometimes called Rayleigh or Love waves.)

The epicenter is the point on Earth's surface that lies directly over the focus, or point of origin, of the earthquake. Effects of an earthquake are greatest in the zone around the epicenter.

Each number on the Richter scale represents an intensity ten times greater than the previous number. To date the highest magnitude detected registered 8.9 on the Richter scale.

The Mercalli scale describes the damage in the following way:
I.   Detected by seismographs and only very sensitive people.
II.   Large buildings move slightly.
III.   Vibrations like those of a truck passing are felt.
IV.   Dishes, windows, etc., vibrate.
V.   Most feel it; some objects fall and break.
VI.   Heavy furniture moves, trees shake, and windows break.
VII.   Poorly-constructed buildings are damaged.

VIII. Slight damage to well-constructed buildings.
IX. Underground pipes break; the ground cracks.
X. Railway tracks bend; landslides occur.
XI. Bridges collapse.
XII. Buildings are totally destroyed.

The Richter scale is logarithmic. Each increase of one represents an intensity ten times greater than the previous number.

Shaken-up Vocabulary (Page 29)

1. seismic   2. fault   3. Richter   4. waves   5. damage   6. tsunami   7. geology

The letters over the numbered lines form the word "seismology."

Other Geologic Phenomena (Page 30)

A fumerole is a hole in a volcanic area from which hot steam and gases arise. In some areas the steam is harnessed to generate electricity.

A hot spring is water which has been made hot as a result of contact with magma; the magma, although near the surface, has not burst through. When a hot spring has a crooked, narrow vent, the hot spring becomes a geyser. A geyser emits water at intervals. Old Faithful, the most famous geyser, is in Yellowstone National Park in Wyoming.

The use of the term "tidal wave" for tsunami is a misnomer because it has nothing to do with tides. Tsunamis result from underwater seismic and volcanic activity.

When in open ocean, tsunamis travel at very high speeds and have wavelengths up to several hundred feet; however, they may have wave heights less than three feet. Because they are so shallow, they go unnoticed by ships that pass over them. As they near the shallow water of the coast, the tsunamis slow down. This causes their lengths to shorten and their heights to rise. Some rise to 100 feet (30 meters). Tsunamis often cause a great deal of damage. Usually, however, seismologists monitoring earthquake activity can warn people of their approach in time to save lives if not property.

When part of a hillslope weakens and is unable to support its own weight, a landslide occurs. A landslide is the movement of rocks, soil, and vegetation.

When large amounts of snow and ice fall suddenly and rapidly down a mountain, it is called an avalanche.

Plate Tectonics (Page 31)

Scientists believe that about 200 million years ago all the continents were joined into one supercontinent, which is called Pangaea.

A rift valley is a long, narrow depression in the earth's surface formed when the land sinks between two fairly parallel faults.

The continents are surrounded by a gently sloping, submerged plain. These plains are called continental shelves. They are exploited for their petroleum, commercial sand and gravel, and fishery resources. The continental shelves are also sites of waste dumping.

Geologic Fact or Opinion (Page 32)

1. F   2. O   3. F   4. O   5. O   6. F   7. O   8. F   9. F   10. F   11. O   12. F   13. O   14. F   15. O

Geologic Syllogisms (Page 33)

1. V   2. I   3. V   4. I   5. I   6. I   7. V

An Ancient Myth (Page 34)

A = Y	E = C	I = G	M = K	Q = O	U = S	Y = W
B = Z	F = D	J = H	N = L	R = P	V = T	Z = X
C = A	G = E	K = I	O = M	S = Q	W = U	
D = B	H = F	L = J	P = N	T = R	X = V	

If students are having trouble deciphering the code, you might want to suggest the following:

1. Think about what the title might be. This will help you decode a few letters.
2. Look at the small words, especially one-letter words.
3. Look at punctuation marks (apostrophes, for example) for clues.

An Ancient Myth

In the Mediterranean Sea is a group of volcanic mountains named the Lipari Islands. The most southern of these mountain islands is called Vulcano. Vulcan was also the name of the ancient Romans' god of fire.

Eventually, the Romans began to identify Vulcan with the Greek god Hephaestus. Hephaestus was god of the forge. Hephaestus, or Vulcan, was often portrayed as a blacksmith, laboring over a hot fire and fashioning beautiful tools and ornaments. When fire came out of their mountains, the ancient Romans thought it was sparks from Vulcan's furnace. Today we still call these mountains "volcanoes" after the ancient god of fire.

Geology Scrambled Words (Page 36)

1. earthquake 3. igneous 5. lava 7. erosion 9. crust 11. avalanche 13. geyser
2. volcano 4. eruption 6. magma 8. mantle 10. mountain 12. landslide 14. geologist

Crossword Puzzle (Page 39)

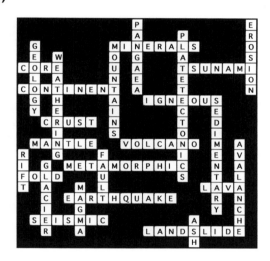

What's the Question (Pages 40–42))

Answers are given on page 40.

Bibliography

Blobaum, Cindy. *Geology Rocks: 50 Hands-on Activities to Explore the Earth.* Charlotte, Vermont: Williamson Publishing, 1999.

Dudman, John. *The Violent Earth: Volcano.* New York: Thomson Learning, 1993.

Rhodes, Frank H. T. *A Golden Guide: Geology.* New York: Western Publishing Company, 1972.

Ruiz, Andres Llamas. *Volcanoes and Earthquakes.* New York: Sterling Publishing Company, Inc., 1997.

Staedter, Tracy. *Reader's Digest Pathfinders: Rocks and Minerals.* Pleasantville, New York: Reader's Digest Children's Publishing, Co., 1999.

Stark, Rebecca. *Create a Center about Earthquakes and Volcanoes.* Hawthorne, New Jersey: Educational Impressions, Inc., 1984.

Walker, Sally M. *Earthquakes.* Minneapolis: Carolrhoda Books, 1996.

Wyckoff, Jerome. *The Story of Geology.* New York: Golden Press, 1976.

Zim, Herbert S. *A Golden Guide: Geology.* New York: Western Publishing Company, 1985.

Printed in Great Britain
by Amazon